The Fantastic Book of Chemistry Jokes: For Everyone, Not Just Chemists: Author's Edition

UNCONVENTIONAL PUBLISHING

www.unconventionalpublishing.com.au

Shane Van

This book is dedicated to two of my good friends, Sandy Maxwell and Tracey Oades.
Sandy has always supported and encouraged me in writing these books; in fact, she even beat me in purchasing the first book I wrote.
Tracey has always been a caring and dear friend, offering support and advice, even checking in on me after months of not seeing me in person.
Thank you both.

ISBN – **978-0-6455601-1-4** Hardback

Disclaimer: This book is a work of fiction and is not to be taken seriously; there is nothing in this book that should be taken as fact, especially scientific fact.

Table of Contents

Preface..7
The Avogadro..11
Simple Chemistry 17
Famous Chemists..27
Physical Chemistry...33
Evil Chemistry 43
Working Chemistry..53
Drunken Chemistry..65
Bad Education...71
Biochemistry...81
Chemistry Relationships 87
The Realistic Glossary.................................101

Preface

Unconventional Publishing proudly presents *The Fantastic Book of Chemistry Jokes: Author's Edition.* This book contains all the jokes that should have been in the first edition. After releasing the first edition, Shane Van and his readers have been persistent in requesting even more chemistry jokes, resulting in the 'Author's Edition' with even more of those funny jokes we love so much!

The original book did not match the size or have the content of the other books in the *Fantastic Book of Jokes* series. The Author's Edition, now has the necessary changes and fun additions that were needed to keep the content in line with the other books of the series.

Nothing in this book should be taken as fact and is only here to put a smile on your face; no part of this book is meant to offend. If it does in any way offend you, you can always put it down and move on to something else.

This book is a part of a series of books that will challenge all the sciences and everyday professions. The series will cover everything from chemists to drummers. Unconventional Publishing is a company that aims to produce books that differ from the norms. Unconventional Publishing aims to satisfy society's unique and different tastes, to delve deep into the realms of science academia, and push the boundaries on what is realistically achievable, and to bring knowledge and evoke discussions into everyday life.

The Avogadro

What was Avogadro arrested for?
He had a very peculiar obsession with moles.

How many pieces of fruit does it take to make guacamole?
An Avogadro number.

"I wish I was Avogadro; then I would already know your number."

What would you have if you had a mole of avocadoes?
A guacamole.

Why does Avogadro play golf on weekends?
He wants a mole in one.

A mole of moles would weigh so much that they would collapse in on themselves and create a black mole.

A mole walks into a bar.
"Sorry, we can't serve you tonight," says the bartender.
"Why not?!!" growls the mole.
"We have a capacity of 6.00×10^{22} and can't accept one mole."

What is the concentration of one tooth in a glass of water?
1M (1 Molar).

What happened when caesium downloaded Tinder?
He got Avogadro's number.

Let me mix my acid with your base and I'll leave behind 100 moles of my salt mixed with water in you.

Chemists do it with moles.

If a mole of moles were digging a mole of mole holes, what would you see?
A mole of molasses.
What is Avogadro's favorite game?
Whack-a-Mole.

What is the quickest way to make a 24-molar solution?
Put your false teeth in some water.

Why do chemists go to jail after mixing 1 mol of alkane with 1 mol of carboxyl acid?
They created 1-mol-ester.

Avogadro used to be a maths teacher. He used to teach his students mole-tiplication.

Why couldn't Avogadro get a date?
His number was too long.

Why was Avogadro executed?
He was a mole.

Why did Avogadro like Cindy Crawford?
She was his favorite super mole.

People shouldn't make jokes about Avogadro. They are molality wrong.

What was Avogadro's favorite fruit?
Watermole-ns.

What did one mole say to another mole?
I think we have great chemistry together.

I have two words to describe the size of Avogadro's number.
Holy-moley.

What was Avogadro's favorite movie?
The Green Mole.

What kind of exam do chemistry students like to take?
Mole-tiple choice.

What does Avogadro mix in his hot chocolate?
Marsh mole-lows.

They say when Avogadro was born, he broke the moled.

Simple Chemistry

How do you stop your atoms from running away?
You keep an ion it.

Be like a proton and stay positive

What is Santa Claus's favorite element?
Holmium.

Which group of elements are religious?
The halo-gans.

A new Marvel movie is coming out; it has Iron Man and the Silver Surfer teaming up.
It's called 'The Alloys'.

Who is Iron Man attracted to?
Any Fe-male.

What is Iron Man's favorite ride?
The ferrous wheel.

What is a chemist's favorite dog?
A Laboratory Retriever.

Three siblings were trying to change the world with recycling. Their names were Poly, Ethel and Ian.

What do you call a boat going through the ocean?
A sodium chlo-ride.

What is uranium 235's favorite PlayStation game?
Half-Life.

Why are noble gases so up themselves?
No matter how you interact with them, there is no reaction.

What do you get if you replace the carbon in a benzene ring with iron?
A ferrous wheel.

If H_2O is the symbol for water, what is H_2O_4?
Drinking, swimming, bathing, showering……… etc.

Did you know that life is an alloy?
Lithium and iron; LiFe.

What elements do you need for life?
Lithium and iron

Why do cats hate water?

They are made of iron, lithium and neon; (FeLiNe).

What does an Italian define as a catalyst?

1 x bag of dry cat food

12 x sachets of wet cat food

1 box of kitty tray liners

2 packs of kitty litter

What did the acid say to the base during titration?

Let's meet at the endpoint.

A bear in India and a bear in Canada both fall into a river; which one dissolves faster?

The one in Canada because it is more polar.

It takes alkynes of people to make a world.

My wealthy great aunt died; she left me all of the antimony.

What did the Japanese chemist say when he mixed nitrogen and sodium?

NaNi.

What kind of element stains tongues?

Tungsten.

What breed of cat lives on the streets?
A-tom cat.

I hate it, I was laying there trying to sleep and I can feel something krypton me.

What is the chemical formula for coffee?
$CoFe_2$.
Why do sheep need barium and hydrogen in their diet?
BaH.

Which alloys are used to cut things?
Potassium, nickel and iron; KNiFe.

What is the empirical formula for seawater?
CH_2O.

Did you know that sodium hydroxide is a liquid out of solution?
Well, that was a lye.

What's the coolest element around?
Gold because its Au-some.

What is the code name for a MI6's Eskim spy?
Bond, Polar Bond.

Why does mincemeat provide the smallest amount of energy?
It's in its ground state.

What is the pH of Vietnamse soup?
pH0.

Mom would often tell us that dinner would be radon 15 mins.

When I heard how many electrons it takes to fill the first orbital of oxygen, I was like :O

16 Sodiums walk into a bar, Batman then follows.
Na Na Na Na Na Na Na Na Na Na Na Na Na Na Na Na Batman.

If you're not part of the solution you're a part of the precipitate.

Greek dressing is only good fresh. Oil and vinegar are not a permanent solution.

Did you hear they are going to make a movie about the periodic table?
They are going to call it the 'Atoms Family'

Why is citric acid a lot more popular than water?
Citric acid is buffer.

Why wasn't carbon talking to hydrogen?
He was mad atom.

What is the difference between a seal and a sea lion?
1 electron.

Which fish has only 2 sodium atoms?
TwoNa.

How does sulphur text oxygen?
It uses a sulphone.

Noble gases walk into a bar. Nobody reacts.

Carbon and hydrogen went on a date, they really bonded.

If the Queen farts, is it a noble gas?

Chemistry jokes are sodium funny; I slap my neon them.

Never be negative, people will think you're an electron.

Which is the funniest gas?
Double helium, HeHe

What is iodine's and caesium's favourite TV show?
CsI.

What is their favourite show at the beach?
CsI Miami.

Did you hear about the joke with cobalt, radon and yttrium?
It was CoRnY

I took a trip to the beach. On the bus I asked the guy next to me, “Do you see any sodium?”
He answers, “Na”, as he was getting of the bus.
I asked this lady walking her dog, “Have you seen any Sodium Hydride?”
“NaH”, she answered.
I was about to give up when I spotted a surfer,
“Have you seen any Sodium Hypobromite?”
He stared straight into my eyes “NaBrO.”

What is the viscosity of milk?
Mu.

What are Lady Gaga's favourite elements?
Radium, arsenic, radon, molybdenum, gallium, oxygen, lanthanum. She never stops singing about them.
Ra Ra Ah Ah Ah, Ro Ma Ro Ma Ma, Ga Ga O La La

What does Dolly Parton put in her pool?
Chlorine, chlorine, chlorine.

What does sodium, nobelium, chlorine and uranium have in common?
Na, No ClU.

What does the sound of an organic train make?
CH_3COOH CH_3COOH.

I don't care what anyone else thinks, silicon breasts are the most natural things in the world. After all, they are organic and make up most of the Earth's crust.

My wife got angry with me at breakfast when I told her that her sister was more attractive. As I turned around, she hit me with a jar of honey, it was a very viscous blow.

What does a non-newtonian fluid and a masochist have in common?
They both get hard when you punch them.

I bought some engine oil for my boat. It was too thick, so I decided to thin it out with some methylated spirits. It then became too thin, so I added more oil. It went too thick again. I kept putting oil or methylated spirits in to balance it. I just got stuck in a viscous cycle.

Which famous author also worked at a mercury mine?
HG wells.

MARIE CURIE

Famous Chemists

Marie and Pierre Curie have to be the most famous scientists ever. They are just radiant.

What is the difference between Mariah Carey and Marie Curie?
One of them glitters while the other glows.

What was Marie Curie's favourite movie?
It's a wonderful half-life.

Little known fact….. but during the war Marie and Pierre also moonlighted as assassins using certain heavy metals. A lot of people were killed by mer-curie poisoning.

During his time, Alfred Nobel was world renowned. He was a scientist, engineer, invented dynamite and also created the Nobel Prize. He just simply blew all his competition away.

Not many people realise, but Cher has done quite a lot of charity work, however she has yet to win a Nobel Peace Prize. I suppose no one wants another Cher-Nobel.

You have heard about Boyle's Law, but have you heard about Cole's Law?
It's thinly sliced cabbage.

I tried explaining quantum physics to a friend.
They just found it a Bohr.

What happens to elements when they lose all their energy and electrons?
They become Bohred.

I would talk about the atomic structure all day long, but I don't want to Bohr you.

I hate people who tell chemistry jokes at parties. I usually tell those Mendeleev.

There was a chemistry symposium, and all the famous scientists were invited. The following was their responses:
Socrates: "I will have to think about it."
Robert Boyle: "This is putting too much pressure on me."
Marie and Pierre Curie: "We are so excited that we are just radiating with joy."
Archimedes: "This is making my heart buoyant."

John Dalton: "I will model my excitement."
Jacob Berzelius: "Expect a long weight for my reply"
Niels Bohr: "If I go, I won't"
Lawrence Bragg: "I will have to consult my crystal lattices."
Robert Bunsen: "I just want to watch things burn."
Dmitri Mendeleev: "I will organise the seating arrangements."

Pascal walks into a bar; the air pressure was too high, so he left.

Niels Bohr and Einstein were having a discussion about the differences between the theory of relativity and quantum mechanics.
"So, let me understand this correctly, hypothetically if I was having sex with my girlfriend and I thrust at the speed of light, would my penis gain infinite mass?" asked Bohr.
"No, you wouldn't but you would definitely create a black hole." Einstein replied.

I was invited to a wedding in a Faraday cage. The ceremony was beautiful but there was no reception.

I would like to talk to you about the relationship between light and magnetism. However, it will keep us talking Faraday or two.

Physical Chemistry

Two eskimos are in a kayak trying to hunt seals when they harpoon a massive one and it drags them quite a distance from the shore. The current then takes them further out and as much as they try they simply can't paddle back. Eventual it turns to night and the temperature becomes freezing. The two eskimos are left with no choice but to try and start a fire in their kayak to try and keep warm. Sure, enough the kayak sinks. It just goes to show that you can't have your kayak and heat it to.

I once dated a girl that spontaneously combusted, she was pretty hot.

Why do hot girls always date cool guys?
To maintain thermal equilibrium.

My girlfriend is like the standard temperature of a molecule. Meaning that she doesn't exist.

England is currently experiencing a heat wave of temperatures around 40°C. Not cool.

I keep telling jokes about temperature. I am slowly losing my friends by degrees.

Did you know that there is an optimum temperature for sex?
It's usually either too hot or too cold, but never the right fucken temperature.

Why do Chinese cars have rear heated windows?
To keep your hands warm when you push them

The official new SI outdoor temperature activity scale:
12^{0}C: Australians are now freezing, putting jumpers, beanies and thermal underwear on.
6^{0}C: The Scandinavians are now wearing shorts and t-shirts and are outside in the summer gardening.
5^{0}C: British cars won't start.
0^{0}C: Distilled water freezes.
- 1^{0}C: Charities begin to supply blankets to homeless people and start running ads on tv, which cost more than the blankets. Teenagers everywhere start pretending they can smoke due to their breath creating mist. Scandinavian's go out for ice cream.
-5^{0}C: Your cat will climb under the covers with you.
-6^{0}C: The Scandinavians, are starting to cook indoors again.

- 8^0C: Ice lakes are hard enough to go skating on. The Scandinavians are now cutting holes in the ice so they can go swimming.

-12^0C: Everyone is putting pictures of the Bahamas or the Maldives in their office and dream of escaping the cold.

-14^0C: Residents in Oslo start putting their heaters on.

-20^0C: German cars are now having trouble starting.

-21°C: Your cat will now want to cuddle into your pyjamas.

-25 ^{0}C: The government issues health warnings of the danger of kissing in public at these temperatures.

-28^0C: The Scandinavian hockey team start their winter training.

-30 ^{0}C: Japanese cars will no longer start. The Scandinavians will now start buttoning up around their neck.

-35 ^{0}C: People will start migration towards the equator.

-39 ^{0}C: Mercury will freeze, most household thermometers will no longer work.

-44 ^{0}C: The car will now try to cuddle in your bed.

-46 ^{0}C: The Scandinavians will now start wearing jumpers.

-50 ^{0}C: The seals and polar bears are now all migrating south towards Mexico. The Scandinavians are now starting to put on beanies.

-52 ^{0}C: All walruses and seals swim south from Greenland and migrate to England.
-60 ^{0}C: Santa closes down his workshop and packs up all his elves and flies south. The Scandinavians start wearing fur coats.
-273.13 ^{0}C: This is absolute freezing. The vibration of all molecules stop at this temperature. The Scandinavians will now admit they feel the cold.

When you do experiments at home involving pressure and heat, one particular experiment involves lighting a matchstick and putting it in a beer bottle. Then, after placing your testicles on top of the beer bottle, eventually the bottle will have a vacuum and your testicles will get sucked inside. If you have tried the reverse of this experiment. Please let me know as soon as possible. In fact, as fast as you can, they are turning blue.

Scientists have just completed a study and found out that hydrostatic pressure from scuba diving is more harmful than previously thought. I think deep down we already knew that.

Why can't flat earthers calculate the volume of the earth?
They have a rounding error.

A chemist, mathematician and an engineer are asked to measure the volume of a volleyball. The mathematician uses a piece of string to measure the circumference and from there, its volume. The chemist pushes the ball into a bucket of water and measures how much water it displaces. The engineer looks at the serial number then searches a catalogue.

So, the nickname of Las Vegas is Sin City, but do you know what Den City is?
Mass over volume.

I heard steel got into a heated argument once, well it makes sense….. it is a tempered metal.
What do you call the amount of heat energy found in traditional Japanese dishes comprising of meat, seafood and vegetables which have been lightly battered and then deep fried?
Tempura-ture.

Did you hear about the bodybuilding chemist?
They called him Kelvin because he was an absolute unit.

Why don't the Jedi use the Kelvin scale?
Only Siths deal in absolutes.

Did you hear about the student that wrote his thesis on heat?
He was a huge fan of the topic.

When doing a chemistry test on exothermic and endothermic reactions I mixed up units for temperature. I got an absolute 0.

Did you hear about the chemist who got a tattoo of a thermos?
They now call him Mr. Thermostat.

I always set my thermostat to 70^0F. That's because it never getter hotter than 69 in my bed!

What is the male version of a 'Karen'?
Kelvin because he is an absolute zero.

What underwear do scientists wear?
Kelvin Klein.

What do you call an ocean that prefers its temperature ~~too~~ between 23.9 and 24.1?
The Specific Ocean.

What do you call an aquarium filled with carbon monoxide?
Oxygen de-fish-init.

Why do hot air balloons use jet burners and not steam?
That's because fireflies and waterfalls.

In my calculations I accidentally got Fahrenheit and millilitres confused. FmL

Endothermic reactions are the coolest but exothermic reactions are the nicest. They make me all warm on the inside.

Two chemists were arguing what the state of water is at -1^{0}C. One of them had a solid argument.

What is the fastest liquid in the world?
Milk: It's past-ur-eyes before you even blink

What is the liquid inside the battery of an iPad?
Apple juice.

What planet has gas, liquid and a solid at the same time?
Uranus.

For the past week my partner has been arguing about how hot our water system is. Today it finally boiled over

A chemist was recently doing a study across both the chemistry department and the psychology department. He discovered that a humans' understanding of jokes can be modelled using fluid mechanics. Let that sink in!

Evil Chemistry

Why is chemistry the meanest of all sciences?
Because it is constantly pushing electrons around.

Did you hear about the ammonia factory that exploded?
The business went insolvent.

The free radicals have revolutionized chemistry.

He was a natural-born *lead*er. Too bad it poisoned him.

Remember, if anyone in chemistry is causing you problems, bleach is also a solution.

Little Johnny was a chemist.
Little Johnny is no more.
What he thought was H_20 was H_2SO_4.

Your momma's so ugly, even carbon won't bond to her.

Did you hear the latest about nitric oxide?
NO!

Sure, hugs are great, but C_4H_7OH (butanal) is way more fun.

What do you call it when a-mean-old (amino) acid turns evil?
Diabolical acid.

Why do atoms lie?
They can't help it; they make up everything.

What did one atom say to the other?
Stop overreacting.

Which is the loneliest element?
The element of surprise, it isn't allowed near the periodic table.

When is it not suspicious to ask for chloroform?
When you work in an organics lab or a morgue.

How to get fired as a chemist:
Give a colleague a beaker of liquid nitrogen and asked, "Does this taste funny to you?"
Show up with a 55-gallon drum of fertilizer and ask where the nearest government building is.
Constantly write 3 potassium as KKK
Every time it was quiet they keep yelling out, "my eyes, my eyes they burn!"
Deny the existence of chemicals.

Casually walking around the lab and urinating into beakers.
When someone's pouring sulfuric acid, sneak up behind them with a blown up paper and burst it.

My partner only eats sodium and silver.
NAg NAg NAg.

I feared writing these chemistry puns, that I was going to get a volatile reaction.

One anion's trash is another cation's treasure

What do you get if you cross a pig and dolphin? A visit from the Animal Ethics Committee and a ban from all further science experiments.

I had to write a 1000-word essay for chemistry on acid. First the paper changed into rainbows, then the paper shot out rainbows before I melted into the floor

Why does the air force put acid in bombs?
To neutralise bases.

What are cations afraid of?
Dog-ions.

Why do feminists hate chemistry?
Because there is no shelium.

Two marijuana pharmacies were having a dispute. They changed locations many times but were unable to increase any sales. They reached a hash equilibrium.

What professions makes the best conductors? The police, copper is a good conductor.

How do uranium oxide particles go to the toilet?
They uranate.

I keep being told that chemistry isn't a solution, but why would I listen to people who failed chemistry.

Oxygen is the most infuriating element to work with. I ask it if it likes to work with a methyl group or ethyl group and it keeps telling me, "Ether"

Which elements are the badasses?
Phosphorus, uranium, nitrogen potassium and sulphur they are the PUNKS of the periodic table.

What do you call it if someone throws sodium chloride at you?
Assault.

Why do protons hate each other?
They find each other repulsive.

Where did bad elements hang out?
In the (fume) hood.

What did you get when Richard Nixon breathed?
Oxymoron.

What did the university student write as an answer to the question, "Describe hydrophobic"?
The fear of her water bill.

The laboratory smelt like rotten eggs all day, the undergrads were all sulphuring.

You would think a metal and non-metal bonding would mean they are being friendly; they are actually stealing each other's electrons.
Ionic isn't it.

The police pulled over a man in Louisiana and searched his car. They found sodium chloride and a 9v in the car. He was taken downtown and charged for battery and assault.

Heisenberg is speeding along the back roads of Berlin when he gets stopped by the police. Furious, the police ask, "Do you know how fast you were going?" Heisenberg says, "Not at all but I know where I am"

What fruit contains barium and two sodiums?
BaNaNa.

What is a pirate's favourite element?
Aaarrrgon

What is a pirate's favourite element?
If they answer *Argon, say* "Why Argon? It's gold stupid!"

If a fire hydrant has H_2O on the inside, what does it have on the outside?
K_9P

Oxygen, hydrogen, sulphur sodium and phosphorus walk out of chemistry class, "OH SNaP."

Chemistry is a girl because it has a periodic system.

So, I’m dumping acid on kids, throwing radioactive spiders at children, killing rich parents and pouring toxic waste on janitors.
I have yet to create a superhero, but it needs to happen soon, I really need to be stopped.

Throwing acid is always bad in someone’s eyes.

Someone spilt sulfuric acid on my face, I wasn’t concerned as I wasn’t going to see the doctor.

There was this guy in town threatening to cover passers-by in acid. The police were able to neutralise him.

What is the difference between chrome and chromium?
Chrome can heat my laptop to 2000^0C and it will be frozen at the same time.

If you heat your solid-state drive to a gaseous state drive, do you end up with cloud storage?

When I was younger, I was completely obsessed with plasma. Lucky it was just a phase.

What do you get when you mix Russian raspberries with lighter fluid?
Rasp-butane

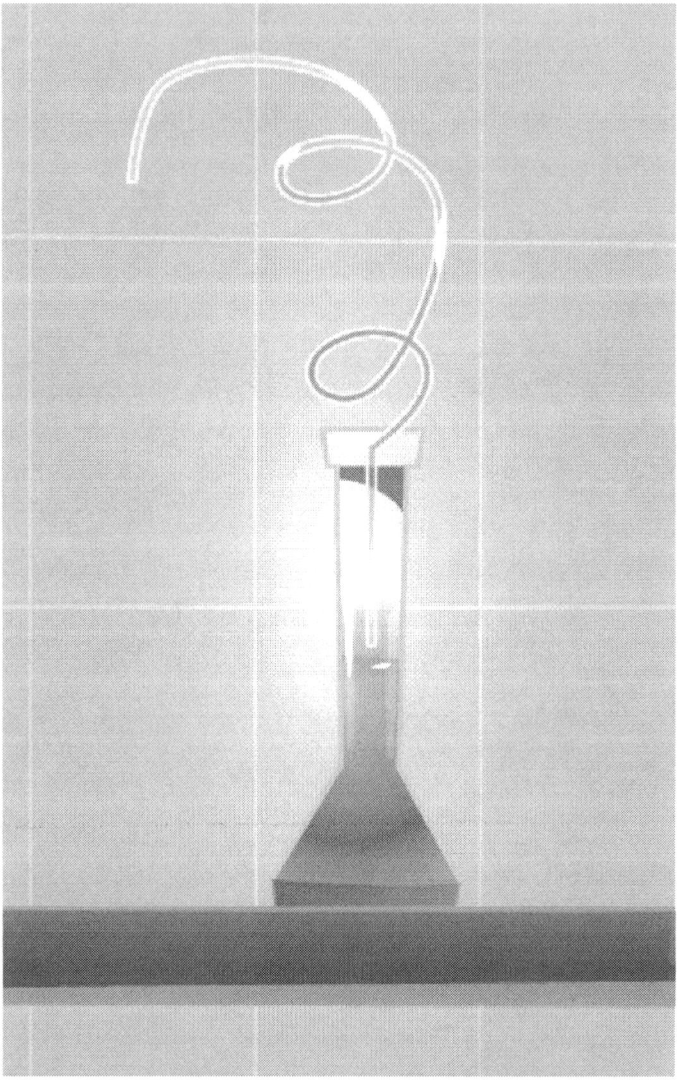

Working Chemistry

Organic chemists are always running into alkynes of problems.

Albert Einstein only lived on a diet of germanium, nickel, uranium, and sulfur because he was a GeNiUS.

Chemical reactions can blow your mind. No, seriously, they can literally make your brain melt.

A pharmacist walks back in from his break. He sees this guy with a pained look on his face; he turns to his assistant and says, "What's up with this guy?"
"Oh, he came in for some cough syrup; since we didn't have any, I sold him some laxatives."
"Why the hell did you do that????"
"Take a look at his face; he is way too scared to cough now."

A guy walks into a pharmacist and asks, "What is good for disinfecting cuts?"
The lady behind the counter says, "Ammonia cleaner."
"I'm sorry, I thought you worked here."

How do doctors treat their chemist patients?
They try to helium then curium; if that doesn't work, then they barium.

How good was the book on helium?
The chemist couldn't put it down.

Why did the lab assistants' pants keep falling down?
They had no acetol.

How many chemists does it take to wash a beaker?
None - that is what lab assistants are for.

In class, a student got some dichromate in their mouth and got sick.
What can you say? Oxidants happen.
How many analytical chemists does it take to change a light bulb?
Only one, but he'll replace it three times, measure the lumen's, plot a straight line, calculate errors, and extrapolate to zero concentration.

How do you know that a chemist is using the bathroom?
He washes his hands *before* he uses the toilet.

A chemist's last words;

- And now the taste testing.
- Is this meant to get hot?
- Trust me – I'm a chemist.
- Don't confuse these 2 beakers.
- Something isn't quite right here.
- Yes, the Bunsen is off.
- Where is the label for this bottle?
- Now we remove the protective shield.
- It should now be at a constant 24^0, Ummm, 25^0, 26^0….
- What is that smell?
- What test tube did you give me?????
- Just tap the edge like so.
- Phew, I am going for a smoke.

A chemist, physicist and a statistician go hunting in the woods. A giant buck leaps in front of them. The chemist quickly gets off a pot shot and misses by exactly 3 ft to the left. The physicist gets on one knee, takes careful aim, and just as he shoots, he slips and misses by exactly 3 ft to the right.
The statistician screams, "Eureka, we got him!"

Organic chemistry, the only profession where asking for chloroform is not suspicious.

Did you hear about the chemist who discovered a new element?

He won a nobelium.

A chemist goes to a pharmacy and asks the pharmacist, "I need some 4-hydroxyphenyl or acetaminophen."

The pharmacist looks at him and says, "Do you mean Panadol or Tylenol?"

"Yes, that's it; sometimes I forget the names."

Two scientists are studying electron movement in orbital energy levels. The first one says to the second, "I don't know about you, but I feel as if there is a lot of negativity in the room."

A chemical substance is;

What an organic chemist turns into foul odor.

What an analytical chemist creates a method for.

What a physical chemist finds out at what temperature it combusts.

What a biochemist turns it into a helix.

What a chemical engineer turns into a profit.

Let's be honest; a chef is just a gastronomical chemist.

An optimist sees the glass half full.
A pessimist sees the glass half empty.
A chemist sees it 100% full but with 50% gas.

Why are chemists so great at maths?
They hold all the solutions.

Two chemists are working when one of them has an experiment fail, the second one says "Ytterbium."

Why do chemists love to work with sodium hydroxide?
It's pretty basic.

How do you tell the difference between a chemist and a dock worker?
Ask them to pronounce unionised.
A man using Apple maps walks into a lab,
……. or a shoe store or a post office.
Why could the chemist hold liquid helium?
Because he was as cool as they come.

A biologist, a physicist and a chemist go to the ocean for the first time.

The biologist is amazed and wants to see the seaweed fields and study its flora and fauna. Running straight into the ocean the riptide promptly takes him away and he drowns.
The physicist was mesmerised by the motion of the waves, wanting to know more about fluid dynamics and wave motion the physicist walked straight into the ocean and also drowned.
The chemist took his notebook out and wrote: *Biologists and physicists are both water soluble*

Why do chemists like working in the lab?
They are in their element.

Two chemists, John and Sandy were working in a lab, when one of them wanted to reduce a ketone without affecting the adjacent aldehyde.
"Why don't you treat the aldehyde with ethylene glycol to form a cyclic acetal. That way you protect the aldehyde from reduction" mentioned Sandy.
"Humph" said John and continued with his work ignoring the advice.
Of course, the reaction reduced both the ketone and the aldehyde, John got very upset.
"Well, if you like it, then you should've put a ring on it" snapped Sandy.

A chemist dared the lab assistant to sit in a freezer set at -273.15^{0}C. He was 0k.

What did the hippy scientist study?
Organic chemistry

How did the hipster chemist burn his hand?
He picked up a beaker before it was cool.

Why do chemists buy shoes with silicon soles? To reduce their carbon footprints.

Top reasons to be a chemist;

- You get exposed to all kinds of toxins and cancer-causing agents.
- Because it's pHun.
- As much pure ethanol as you want.
- Clark Kent-style safety glasses.
- You become an expert at poverty cooking.
- Your diet consists of coffee.
- You spend half the working day answering the question, "Where are the results?"

- Those same clients also want to blame you for global warming, plastics in the oceans and every other environmental disaster.
- And cancer.
- But you do have access to a lot of chemicals and know how to dissolve bodies.
- A physicist, chemist and a statistician are working through the research department, when the see a bin on fire.
- The physicist goes to grab it, “We need to put this in the cool room, that way the temperature will be below the ignition point and it will put itself out.”
- The chemist grabs his arm, “No, we just need to smother it and thus removing any oxygen, in the absence of reactants, the fire can no longer continue.”
- Meanwhile the statistician starts running down the foyer lighting all the bins on fire.
- “But first we need a proper sample size.”

A chemist never dies, they just simply stop reacting.

Did you hear the rumor on the dangerous acid?
Turns out it was a baseless fact.

An organic chemist, an analytical chemist and physical chemist go to the horse races one day.
They decided to pool all their money together and place their bets on a horse but have trouble selecting a horse.
The organic chemist wants to know what each horse is eating and any drugs the horse has been given. The analytical chemist wants to know how much precipitation has occurred and what the soil of the track is like.
The physical chemist started out by hypothesising the horse as a perfect sphere.

I was recently at a chemistry convention. The announcer told us he just invented an acid that ate through everything.
So, I asked, “What does he keep it in?”

All coffee is an acidic solution, except in the staff room, it has a pH of 14. It’s extremely basic.

Which baseball refused to take acid?
Al Kaline

In the not-too-distant future, humanity will have colonized our neighboring planets. Mars has no atmosphere at all and is constantly bombarded with solar radiation. Venus is a giant acidic pressure cooker that melts everything. In comparison to Earth's weather, that is just first world problems.

Drunken Chemistry

When do chemists drink?
Periodically.

At university people do a lot of experiments with drugs, sex, and alcohol but sadly I was in the control group.

Gold walks into a bar, a non-metals bar.
"Au, outside!" yells the bartender.

A photon walks into a motel, the receptionist asks, "Do you need someone to help you with your bags?"
"No, I always travel light."

The reason why chemists drink so much is because they know that alcohol is a solution.

Two chemists go to the bar for a drink. The first one yells to the bartender "A shot of H_2O", the second one yells "A shot of H_2O too".
The second chemist drops dead.
A neutron walks into a bar, and asks "How much for a shot?" The barman replies "For you, no charge."

Xenon walks into a bar for a drink, “We don’t serve the nobles in here” says the bartender. Xenon just stood there and didn’t react.

New Years parties at my house reminded me of a chemistry laboratory. Some people are dropping the base, others are dropping acid and there is a lot of bonding going on.

What do you get if you cross an organic compound in which a carbon atom of an alkyl is bonded to a hydroxyl?
And people say alcohol is never the solution.

I hate playing poker with proteins, all they do is fold.

Two professors in their sixties are discussing their problems in the university’s lunchroom.
“You know, the more I teach the more I feel like I am losing concentration in my classrooms. I was in my biochemistry lecture, and I accidently said orgasm instead of organism” said one professor.

"I know what you mean. I did a very similar thing, the other day I was having breakfast with my wife. I wanted to say, darling can you please pass the sugar. Instead, I accidently let slip, you fucken bitch you are ruining my life."

I took some acid before seeing the optometrist, I passed with flying colors

A chemist was having a drink at a bar and started chatting with a drunk.

"So, what is the most acidic thing you have?"

He asked.

"Oh, it's me phone" answered the drunk.

"I said the most acidic, not the dirtiest."

"Oh, yeah, it's got a *pH* of *one*!"

A chemistry teacher asked their student, "What do you get when you mix acid and a base together?"

The start of a great party was not the correct answer.

When I celebrate my birthday with my partner, we celebrate like an organic compound from an aldehyde group. It always ends in anal.

Why couldn't hemiacetal maintain any serious relationships?

He wasn't very stable and always had alcohol on him.

What does pouring concentrated sulfuric acid down a drain and a Dutch stripper have in common?
They both slowly remove clogs.
What is the difference between farting after eating spicy Indian food and evaporation?
Evaporation is when a liquid becomes a gas.

A wife called her chemist husband who was working late with his team in the lab.
"Hi, look, it's Saturday, I am just wondering when you are going to be home?"
"Sorry, I am just running this experiment and it's all hands on deck" he replies
"What experiment?"
"We've got some derivatives of C_2H_5OH at ambient temperature mixed with a solution composed of H_2O which has CO_2 diffuse throughout it. To cool the mixture, we are adding super cooled, solidified H_2O. We are actually now waiting for some proteins to be delivered so we are also fumigating the area with nicotine vapors. We probably have to repeat the experiment another 4 or 5 times to get a good spread of data.
"Ok honey, try not to work too hard."

Do you make grass white? Do you make grass slippery? Do you make windows wet? Are you a morning person? If so, you maybe dew condensation.

Bad Education

Little Jimmy comes home all happy and excited.
His mum asks him, “How was school, honey?”
“Great mum, today we learnt about explosive chemicals in chemistry class,” Jimmy replied
“Oh, that’s a bit concerning, so what are you learning at school tomorrow?”
“What school?”

A chemistry teacher asks her class, “Ok, class, what is the formula for water?”
Little Robbie raises his hand, “Miss, miss, its H, I, J, K, L, M, N, O.”
“No, Robbie, it's H_2O.”
“That’s what I said.”

Why was the chemist disappointed in her son’s report card?
Because it reminded her of tetrafluoroethylene (2 C’s and 4 F’s).

Study tip for chemistry students.
Don’t drink water when studying; it decreases your concentration

A chemistry professor has a beaker of solvent. He pulls out a $50 note and puts it in the beaker. He turns to the class and asks,
"Will this $50 dissolve?"
Jimmy answers, "No, Sir."
"Excellent, Jimmy. Can you tell me why?"
"Because you're too much of a tight ass to dissolve $50."

A chemistry professor is starting her lecture but is having computer problems. She then asks her class, "Does anyone know how to unfreeze a computer?"
Jimmy puts his hand up nervously, "Um, miss, what is the melting point?"

A chemistry professor had a new group of students. They were awful, talking back, making smartass comments all the time, rarely letting him finish a sentence.
During a practical class, they were all standing around his bench for a demonstration. In the center of the table was a beaker with some orange liquid in it.
"Ok class, to be a good chemist, you have to be very observant and take notes of the smallest things." He said as he dipped his finger in the liquid. He then sucked on a finger.
"Ok class, please do as I did."

So, the students, each stuck a finger in the liquid, then they sucked.
"Ok class, what did you observe?"
"Oh, it's just dyed water; that was pointless." remarked one student.
"No, it tastes slightly salty," said another.
Sure enough, an argument started with the students. The professor then went back to his desk, grinning from ear to ear.
"Sir, why are you smiling?" asked a student.
"Well, if you were observant, you would have noticed that I put my pointer in the liquid but sucked on my pinkie. Meanwhile, you just all tasted my urine."

During chemistry class, the experiment was a group activity, and the students had to weigh and accurately record approximately 20 grams. One student went to borrow a pen to write the result. As soon as he picked it up, another student grabbed the pen and yelled, "Bromine!"
The student just shrugged his shoulders and went back to the experiment. He was watching his classmate weigh the sample. She reached 43 grams.
"Aren't you getting a bit overweight there?", he said
She snapped, "None of your Bismuth!!"

During university finals for chemistry, the very last question given to the students was:

<u>‘Is hell exothermic or endothermic? Discuss using theory.’</u>

Most students spoke of the Molecular Theory of Particles in an excited state; when they have more energy, they expand and become gas or lose energy and become liquids, then solids.
One student took a different approach;
The first constant that is needed is mass, the actual mass of hell. This, unfortunately, is not constant as hell is not a closed system; souls are entering and leaving all the time. Hell, being hell would not allow many souls to escape, so it is safe to assume that the mass is constantly increasing.
How many souls are actually entering hell? If we take into account the world’s religions, the vast majority of them believe in hell so those souls would go to hell. If we also look at the world's population. It is expanding at an exponential rate. The rate of births is exceeding the rate of deaths. Thus, we can hypothesize that the rate of souls entering hell would also increase exponentially as well.

If you look at Boyles Law and the relationships between volume, pressure and temperature, we have to have to decide that either;

Hell's volume is not increasing or increasing slower than the number of souls entering; then, hell must be exothermic. It will keep getting hotter until all hell breaks loose.

Hell's volume is increasing faster than the souls entering it, which would mean an overall energy loss, then hell would be endothermic, and hell would indeed freeze over.

Which is it then?

A fellow student, Diana, had postulated to me in high school that she would only go on a date with me if hell freezes over. This was said to me 4 years ago. However, 2 nights ago, that date did occur. So now we definitely know that hell is endothermic.

On a side note, the existent of God was also proven as by the end of the night, she kept screaming, "Oh God, Oh God, Oh God!"

He received an A+

Which is smarter, a thermometer or a beaker? The thermometer is smarter, the beaker may have graduated but the thermometer has a lot of degrees.

What's the difference between hospitality and chemistry class?
You do not want to lick the spoon after chemistry class.

A chemistry teacher was yelling at his students for not knowing any of the chemical symbols.
"Back in my day I would know all the symbols and their masses."
"Back in your day sir, there was only 20 odd elements."

In a catholic school the priest teaching science says to the class, "Did you know that protons have mass?"
"I didn't know they were catholic," said a student.

Student 1: Do you get zinc sulphate by mixing zinc and sulphate?
Student 2: I Zinc so.

A science teacher asks the class, "I have two liquids, water and butane. What would be a good measurement of liquids?"
"Please sir, I would use litres" shouts out little Jimmy.
"Very good Jimmy, now classify which one is heavier?"
"The water," shouts out Jimmy again, "because butane is a lighter fluid."

In chemistry class the students were given an assignment of studying chemical reactions and recreating one. As the students were mixing various chemicals, some making hydrogen gas bubbles and popping them, others creating endothermic reactions while the teacher was walking around the room helping students. He notices little Jimmy who never studies, about to put 1kg of potassium into water.
"Ah Jimmy, you need to stir the water for 5 mins first" says the teacher.
"Why Sir?" asks Jimmy.
"It will give me a chance to get the hell out of here first."

My chemistry teacher asked me to rank all the bonds. So, I did;
Daniel Craig
Sean Connery
Roger Moore
George Lazenby
Timothy Dalton
David Nevin
Pierce Brosnan

What are the main things you learned in organic chemistry?
How to connect the dots and how to draw hexagons.

What do chemists drink?
Beryllium- Erbium.

A sign outside the science lab reads; "Great day rates but better NO_3^-"

Four university friends had an end of year exam for chemistry on a Monday morning. They were also invited to the Women's Volleyball Grand Finale at another university in another state. Not wanting to miss the opportunity to mix with some ladies, the boys decide to drive six hours across the state, then leave early the next morning so they would have Sunday afternoon to study.

So, the four boys left on a merry old road trip. They got to the game, celebrated at the after party and drank, and kept drinking the night away. Come Sunday the boys were too hungover to drive back, let alone study. They eventually made it back late Sunday night with no time to study.

They hatched a plan; they decided to email their university lecturer and beg for forgiveness. They explained that they had to help a sick aunt move house but when they were coming home, they got a flat tyre and being in a small town, it took 24 hours to get the tyre. Can they please have extension or re-sit. The professor replied and gave them another three days.

Three days later the boys came to sit the exam. The exam was going to be out of a 100 point, and they blitzed the first page, it was simply looking up masses from the periodic table - worth 15 points. Then they turned the page and looked at each other in shock and horror.

The last page read "For 85 points each of you have to say which tyre and all the answers have to be the same"

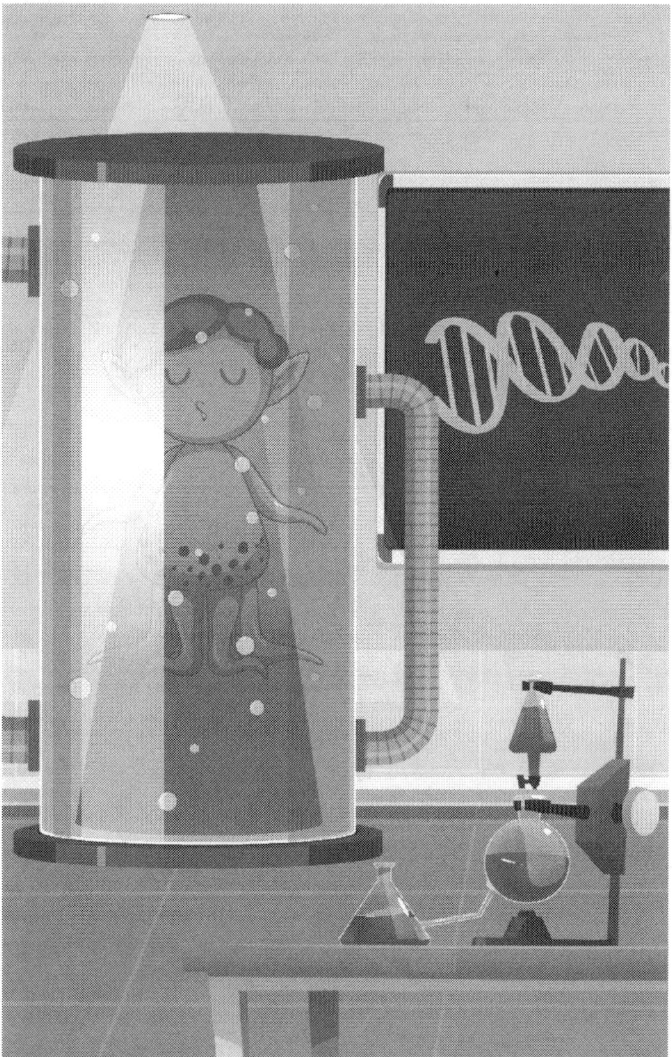

Biochemistry

Where do amino acids go to church?
The cysteine chapel.

I had a friend that was allergic to citric acid,
every time life gave her lemons, she had an anaphylaxis.

Are you a biochemist?
Because your face makes a hormone.

How do you make a protein?
You combine chains of amino acids.
How do you make an enzyme?
You encase the proteins around each other.
How do you make a hormone?
Don't pay her.

How do you make a hormone?
Leave me with your mother for an hour.

How do you make a hormone?
Kick'em in the crotch.

What is the difference between an enzyme and a hormone?
You can't hear an enzyme
Did you hear about the biochemist alien that made contact with humans?
He said, "Amino harm."

Why do amino acids only go to juvenile detention?
They are pro-teens.

What do formic acid and heart-burn pills have in common?
They are both anti-acids.

What does DNA stand for?
National Dyslexia Association.

What sound does a fatty acid make when it sneezes?
A-COOH.

What do you call some amino acids who are just chilling?
Residudes.

A chemist went to see a health and safety occupational therapist about difficulties he was having in the workplace.
"You see doc, I have this irrational fear of hydroxyl groups?"
"Oh??" replied the doctor
"Aaarrrggghhh." screamed the chemist

What are Mexican proteins made of?
Amigo acids

What did the biochemist say when she was asked for the time?
Threonine

An aromatic amine and carbonyl walk into a bar.
The aromatic yells out "Hey, let's get Schiff based."
There was a hare hopping through the meadows, enjoying the smell of spring and the nice cool day. As he was hopping through the meadow, he comes across a lonely zebra who was about to light up a large joint.
"Stop, Mr Zebra stop!!! Don't you know that drugs are bad for you? Put it down and come follow me and let's see what fun we can have." exclaims the hare.
The zebra looked at the hare, then back at the joint. He flicked it into the forest before joining the hare.

The zebra and hare were galloping and hopping together having a great time when they came across an owl. The owl is sitting there with a baggie of heroin.

"Stop, Mr Owl stop!! Don't you know that drugs are bad for you? Put it down and come follow us and let's see what fun we can have." exclaims the hare.

The owl looks at the baggie and then throws it to ground before he flaps down to join them.

The owl, zebra and hare are now flapping, galloping and hopping through the forest all enjoying each other's company when they come across a lion. The lion was about to have a beer.

"Stop, Mr. Lion stop!!! Don't you know that alcohol is bad for you? Put it down and come follow me and let's see what fun we can have." exclaims the hare.

The lion glances sideways at them then back at the beer. With one sharp claw he pops the lid off.

"Stop, Mr. Lion stop!!!" yells the hare.

With a mighty swipe, the lion launches the hare across the meadow straight in a tree, crunch.

"What the FUCK Mr. Lion, He was only trying to help you?" screams the zebra with tears running down his face.

The lion then grabs the zebra by the scruff of his neck.

"I am not a lion, Josh I am Chris, your roommate. You are a 38-year-old man who has dropped acid in the lounge room and has been talking to his fucken hallucinations in the middle of the day. You need to get your shit together!!!!"

There was once a biologist sent into the swamps to study the mating ritual of a rare and endangered green swamp frog. He often noted that during mating the frogs could not remain coupled together long enough for fertilization to occur. He thought it had something to do with the polluted swamp water, so he called in help from the biochemistry department at MIT.

The biochemist took some water samples and some frogs back to the lab and spent weeks testing out different chemical compounds. In the end he finally figured it out. He called in the biologists for a demonstration. There was a tank with a male and female frog. He added some nutrients into the water and then simply added some sodium.

"Why did you add the sodium?" asked the biologist

"That is the most important part, it is because it is monosodium glu-ta-mate."

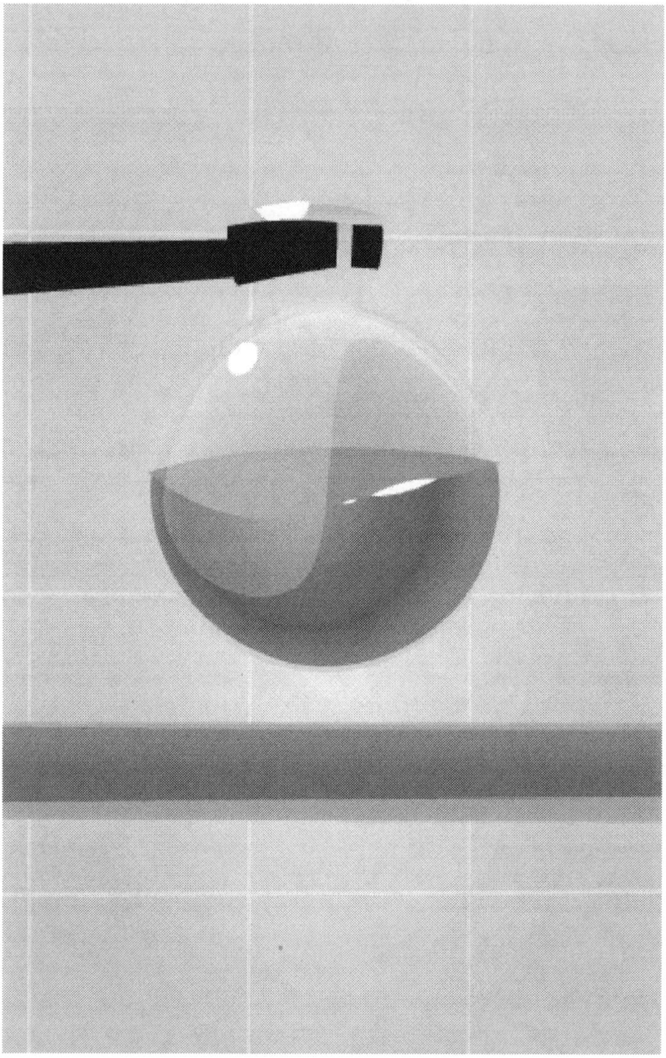

Chemistry Relationships

A university professor asked a female student in his class, “What is a nitrate?”
The girl answered, “Usually $300 an hour and taxi fare.”

Men’s Tinder profiles always say, ‘I am looking for chemistry, the right person and love.’ When will they be honest and say, ‘I am looking for a drug dealer and hookers?’

Babe, you must be exothermic because I am feeling how hot you are!

Babe, you must be endothermic because you are taking all my energy!

Babe, do you eat copper and tellurium for breakfast? Because you’re CuTe.

Babe you must be made of 11 protons because you are sodium fine.
Do you like science? Because tonight you are going to sample my DNA.

Let's study at mine? I have a molecular model kit, so that you can play with my stick and balls.

You make me so hot that I could have a nuclear reactor meltdown.

A reaction between you and me would be exothermic. Let's go collect some data points.

Even the Kelvin scale could not possibly measure how hot you are!

According to the 2nd law of thermodynamics, you need to share your hotness with me.

You must be made of diamonds; you are giving me a hardness of 10.

You must be made of azidoazide azide because if I get you wet, then the reaction will be explosive.
Do you have a solid core of iron? Because you magnetize me.

Babe, you must be made up of fluorine, iodine and neon, because you are so damn FINe.

Babes, did you have copper for dinner? Because I Cu in my bed.

You're a moving electric charge, and I am a magnetic field. So, let's go flux.

What did the liquid chromatography say to the gas chromatography?
Breaking up is hard to do.

Chemists do it on the lab bench periodically.

Why did sodium hydroxide go to the gym?
To become a buffer solution.

My heart is made up of gallium; it melts every time you touch me.

You have a lovely pair of boron, radium and strontium compounds there.

They call me the Higgin Bose particle because I'll be making you scream "Oh God" in the end.

You're on fire, quick get in the safety shower with me.

Are you sure we didn't have a science class together? We do have a lot of chemistry.

Would you like to join my partner and I for a triple bond? They are alkynes of fun.

Babes, you must be an electron; you have some potential.

Hey beautiful, I know my science and you have one significant figure. I would love to subtract your clothes, divide your legs then square root you.

They call me a photon; I will get your electrons to an excited state.

Girl, are you made up of 67 protons? Because you're a Ho.

Author: It's ok, babes. I know you're sick of my chemistry jokes.
Partner: I AM sick of hearing them (they begin to tear up).
Author: Come here and baryon head on my shoulder and cry.

Hey baby, your eyes are full of beryllium, gold and titanium, because they are BeAuTi-Full.

There was a periodic table pool party; oxygen yelled to another oxygen, "Hey, get in here, for every oxygen, there are 2 hydrogens!"

Oxygen and potassium went on a tinder date.
It was OK.

What did the horny chemist say to her colleague?
Compound me.

Hydrogen and helium are walking down the street. Hydrogen stops suddenly and starts yelling, "I've lost an electron, I've lost an electron".
"Are you sure?" says helium.
"I'm positive," says hydrogen.

A woman goes to a pharmacy to buy some condoms.
"Would you like a bag, miss?"
"No thanks," she replies, "He isn't that ugly"

A guy walks into a pharmacy and requests 50 condoms. The cashier says with a smirk, "WOW, you have a busy weekend planned."

“I sure do. I am making a raincoat for my pet snake,” he replies.

Her: I don’t want to hear any more fucking chemistry jokes; I swear I will explode.
Him: Potassium

Carbon and hydrogen went on a date; they really bonded.

A retiree goes to a pharmacist for help with impotence.
“Sure,” says the pharmacist, “we have Viagra, I know it keeps my missus happy”
“Can you get it over the counter?” says the retiree.
“Only if I take 3 in one go.” says the pharmacist.

If you were graphite, I would be an electron, so I could travel through your sheets.

I want to twirl my stirring rod in your beaker.

You must be a carbon sample, because I really want to date you.

This girl gave me arsenic sulphide, I tore that ASS up.

Babe, are you the litmus paper for my acid? Every-time you touch it, it goes red

You make me so hot, Chernobyl looks weak in comparison.

Babes you must have a diet of oxygen and neon, because you are the one.

Do you want to share some electrons and have a stable relationship?

If you were an element, you would be francium, because Germany wants to invade your body.

Our relationship will be more explosive than potassium and water.

I might have a short bond length, but I have a thick electron density.

Solar neutrinos are penetrating you every second, can I join?

Babes you are just like sodium, if you get me wet, I'll explode.

Babes are you made of Zinc?
Because when you're hot, you bond with everything.

Babes, do you want to find out our combined density?
Let's displace the shower water.

Do they call you barium beryllium?
Because you're the BaBe.

Chemists tell you that you are made from 60% oxygen. Biologists say that you're made of 70% water and physicists say you are 99.99% space, but I'll say you are 100% cute.

What do kinky bondage fetish chemists use as safe words?
Nitrogen Monoxide.

What psychological disorder do chromatographers suffer from?
Separation anxiety.

Why did the noble gas cry?
Because his friends argon.

Why does hydrogen oxide bond with everything?
Because they are a HO.

Did you hear about the threesome between oxygen and two hydrogens?
It made me wet.

You think you're a 10?
Well, that would be on the Ph scale because you're BASIC.

A covalent compound was busy yelling at an ionic compound. "Oh really? Can't you ever share."

You must have a phenomenal heat capacity because you are hot.

You must be zinc sulphate and I must be copper sulphate with a salt bridge, because I can feel the electricity between us.

Can I be inside of you and really feel those Van der Waals forces?

Wow you must be a 14 on the pH scale, you're the most basic need in my life.

Your lab or mine?

You must be from the far right of the periodic table, your electro negativity is just pulling me.

You must be a non-volatile particle; you are giving me a rise…. in boiling points.

Babes you don't need any more diet coke, you have a great ASSpartame.

A man walks into a pharmacy and sees a lady working behind the counter.
He nervously asks her, "Do you have a male chemist working here?"
"No," she replies "but I am more than happy to help you"
"It's just, I'm a bit nervous and embarrassed by my condition"
"My dear, my sister and I started this pharmacy 30 years ago, we are in this store day and night. We have given our lives to this store, and we haven't even had time to find ourselves partners. While we have been here, we have come across almost every aliment known to mankind, you have nothing to be ashamed of, I am sure I can help."

The man looks down at the counter and begins muttering, "Well I have this constant erection, it never seems to go down, it's always hard. It's embarrassing I've even lost jobs because of it. Is there anything you can do for me?"
"Oh, well that is something new, let me talk to my sister and see what we can come up with."
So, the lady walks out the back yelling "Sister, sister". After about 10 minutes she comes back to the front counter.
"Ok, the best we can offer, is free food and board, the truck and $200 week spending money."

A couple who were scientists had twins, they called one Jessica and the other Control.

For the entire party I had my iron you.

What is a future girl's best friend?
Carbon

I would dump my girlfriend in sulphuric acid, but the basic bitch would neutralize it.

Damn girl, are you a strand of DNA helicase?
Cause, I see you unzipping my genes

I am oxygen and you're magnesium; I will light you up.

A small piece of metal was living in a test tube and fell in love with a Bunsen burner. "Oh, I just melt for you my dear." It replies,
"Don't, you are just going through a phase", the Bunsen replied

My wife got a bottle of acid thrown at her while we were going out for dinner. I immediately took her straight to hospital.
"Will you still find me attractive?"
"Sure, I will, but maybe wear a mask for a while, before we can get a surgeon for a facial reconstruction. I can also donate some of my skin for a graft if I have to."
"You bastard, the acid only hit my leg."

The Realistic Glossary

A

Absolute temperature: the temperature vodka should be served at
Absolute freezing: same as above but in Siberia
Acid: taken at a party
Acid Rain: when it kicks in
Activation Energy: the amount of coffee needed to start the day
Adams, John Conch: someone who crossed proctology and astronomy
Aerobic: a flying biscuit
Age of Exploration: the time England travelled the world and discovered 'new' places and people that existed for thousands of years
Agronomist: a scientist that likes dirt
Alcohol: hehehe
Alchemy: a 'want to be' type of chemistry mixed with wizardry
Alkynes: all types
Alpha Particle: the particle that gets all the ladies
Asthenosphere: getting into Astheno's pants.
Astrobiology: aliens
Atomic Mass: where atoms pray

Avogadro, Amedeo: an Italian guy, who plucked a large, odd number out of thin air and made all chemists use it for the next thousand years. He also likes avocados

B

Bacon, Roger: a friar from ages ago who brought gunpowder to Europe to pretend to be a magician
Base: what all good music needs
Becquerel, Henri: the other guy who got the award with Marie and Pierre Curie, but no one ever talks about
Berzeluis, Jons Jakob: got famous by electrocuting sick people (scientists got away with everything back then). He also discovered some elements

Big Bang: when everything was created in a loud explosion. It also begs the question, 'Did it even make a noise since no one was there to hear it?'
Biogeochemical: something that couldn't make up its mind
Big Bang: when a chemist loses his virginity, usually at the age of 23
Biogeochemical: something that couldn't make up its mind
Biological Science: a contradiction in terms
Black-Bodied Radiation: a racist radiation

Boyles, Robert: some dude who used to fart in bottles, did figured out pressure volume relationships
Buoyant: opposite of a girl ant
Bunsen Burner: originally invented to be able to brew coffee quickly in the lab, also much less likely to get poisoned than from the work cafeteria
Butyl: alcohol that smells like ass juice

C

Cannizzaro, Stanislao: another Italian guy and Avogadro's publicist
CAI: Computer-Aided Instruction, a way that helps chemistry teachers be lazy
Carbon-14-Dating: when a carbon atom becomes of age and is allowed to bond with other elements
Cation: when a cat irons
Cavendish: a new way to get kids addicted to nicotine
Charge: what happens if Carbon-13-dating occurs
Charles, Jacques-Alexandre-Cesar: a guy that used to light his farts and found a relationship between gas and temperature
Chemical Engineering: stealing the ideas of organic chemists then selling them
Clinical Testing: human guinea pigs

Combustion: what happens when a large-breasted woman wears a very tight shirt
Compound: to make the situation worse
Computer resource: what you add to your budget to buy new toys
Covalent Bond: a love that needs to be kept secret
Covalent Molecule: a ninja molecule
Curie, Marie: woman who won the Nobel prize, in radiation. Played with it so much her house is still a biohazard. She did not get superpowers
Curie, Pierre: first man in history to be completely overshadowed by his wife. Also died of radiation poisoning, did not get superpowers

D

Decay: usually where de beach is
Denitrification: removing of hookers from an urban area
Dense: not smart
Diatomic: dying atoms
Dipole: used to kill atoms
DNA: a chain of acid with a hard name that is used to bring back dinosaurs.
Ductile: a floor tile with ducks on it

E

Eigen Function: when Eigen throws a party
Electrical Charge: what electrons get taken to the police station for
Endothermic: something that sucks the life out of you, financially, spiritually and mentally
Entropy: trying to describe confusion but sounding smart at the same time
Evaporation Allowance: how much undergrads smoke in a year
Excited-state: when the kids hear the ice cream truck drive-by
Exothermic: when you're so hot that your sweat steams
Extrapolate: the only thing analytical chemists do

F

Fahrenheit, Daniel Gabriel: inventor of a scale of temperature that is relatively useless as no scientists and most countries don't use it. Confusing kids in science class for decades
Faraday, Michael: has the love of all conspiracy theory nuts for building cages
Fick, Adolf: great scientist, bad name
First Ionization Energy: the energy you have when you start ironing compared to at the end

First Order Reaction: when Kylo Ren issues an order, and everyone around him, groans at his stupidity
Flame Test: lighting shit up just to see the colours
Fluorescence: glowy stuff, usually poisonous. A friend of mine once drank this and then he was able to predict the future. He said he was going to die then he did
Franklin, Benjamin: All good round guy, built roads and libraries signed the Declaration of Independence and the Constitution, also signed the anti-slavery laws. Invented the human lightning rod by playing with lightning and kites

G

Gas: another name for farts (the most annoying type of jokes and the infernal "pull my finger")
Gay-Lussac, Joseph-Louis: the father of pull my finger
Glucose: too much of it and it will go to your hips
Gordon Neil: good at networking
Ground State: when skydivers hit the ground

H

Haber, Fritz: able to pull nitrogen out of the air. Great for making nitrogen-rich fertilizers, a favorite for farmers and guys who look for directions to government buildings
Half-life: best video game ever
Heat: a movie starring Eddie Murphy
Heterogenous Mixture: a swingers party
Homogenous Mixture: a same sex swingers party

I

Ideal Gas: 'pull my finger.'
Indicator: what inconsiderate asses don't use on roundabouts or turn circles.
Inorganic Chemistry: This is the leftover residue after the organic, physical and analytical chemists have had their fun
Insoluble: acid proof

L

Limiting Factor: what stops things from reaching their potential or how much money you have left in your wallet

Limiting Nutrient: to get around this, athletes use Human Growth Hormone

Line Francis: made up a substance that magically appears in vacuums to extinguish vacuums. Robert Boyle destroyed this idea

Linus Franciscus: same guy as above but changed his name

Liquid: the miracle substance that has been getting people drunk since it was invented

M

Macroscopic: big, big enough to see properly

Magnetic Moment: the moment you see your soul mate

Magnetism: something physical chemists have

Malleability: the bending of metal, the Japanese were masters of this

Manhattan Project: the super-secret project that made the atomic bomb, a not-so-subtle bomb

Metallic bond: bonding over rock music

Microscopic: something so so tiny, you need a machine to see it

Mole: A German scientist trying to get a job in the Manhattan project or a spy

Monomer: 1 mer

Montgolfier, Jacques-Etienne & Montgolfier, Joseph-Michel: these were the first brothers to fly, actually closer to floating then flying

N

Natural Product: something that is grown without pesticides or expansive chemicals but has a sticker slapped on it and charged double for
Net Movement: bait fishing
Neutralization: get neutered
Network: online gaming
Nonpolar Molecular: a molecule that comes from the equator
Nuclear Fusion: the power of the sun. What we have been trying to achieve for the past 80 years.
Nuclear Fission: how atomic bombs work

O

Organic: another reason to put a sticker on something and hike up the price
Organic chemistry: the area of chemistry where you mix one clear substance with another clear substance to create a third clear substance. Hoorah

P

Partition Function: shield built in the lab to protect the supervisor from the under grads accidents
Peptide: a favourite of sports stars
Pharmacology testing: using rabbits instead of people for testing
Phospholipids: a fatty phosphorus
Photosynthesizing: developing film
Physical Chemists: their sole job is blowing stuff up in the name of science
Pigment: what a hog actually means when expressing an idea
Pilot Plant: a small production plant that costs twice as much as a large one
Planks Constant: describes that there will always be some idiot trying to do this on a balcony somewhere
Polar: coming from the Arctic, the bears there are very soluble
Polar Covalent Bond: when two bears who are spies in the Arctic fall in love
Polymers: many Mers
Potential Energy: how much energy is bottled up when your partner says, “I’m fine” or “nothing’s wrong”

Product: what happens when a daddy molecule and a mummy molecule love each other very much and have baby molecules
Priestly, Joseph: credited for discovering oxygen, you know, that gas that every living creature has been breathing from the beginning of time

Q

Quantum Mechanics: a team of people hired to fix Quantum's when they break down
Quantum numbers: a set of points where quantum mechanics look for things

R

Radiation: something that does not give you superpowers
Ramsay, William: discovered the how to find noble gas, allegedly swore like a famous chef
Random Walk: molecules actually behaving like a drunkard when staggering from areas of different concentrations
Rate Equation: to give a grade on how easy equations are to solve
Reactant: an ant that mimics you

Reaction: what happens when a husband comes home, drunk after 'one' drink with work colleagues' and pisses in the sink.
Reaction Rate: how quick they get a black eye from the wife
Repel: what void shields do
Richmann, Georg Wilhelm: tried to be like Franklin, got fried instead
Rutherford, Ernest: co-creator of the video game Half-Life

S

Saponification: this famous reaction was immortalised in the movie Fight Club
Salt: fruits of forbidden love between an acid and a base
Schrodinger, Erwin: crazy cat killer, PETA was definitely not a fan
Scientific Method: the type of testing peers do, to break the hopes and dreams of fellow colleagues
Sediment: what the wife gets when the husband leaves
Solid: hard
Solution: alcohol is the answer
Species: a movie about a super-hot alien trying to have babies

Spectral Lines: the magic lines hidden in light that tell us what stars are made of

Spectrophotometry: a word used to humiliate people like me who have speech impediments

Standard pressure: paying the mortgage and bills on time

Standard temperature: can be either Celsius or Kelvin, not that wank stick of an increment, Fahrenheit

Strong Acid: better than a weak acid

Strong Base: better than a weak base

Surface Tension: Something that chemistry teachers have been using for centuries as a cheap experiment to entertain children

Synthetic: fake

T

Titration: judging at a wet t-shirt competition

Toxicology: testing on rabbits again but getting bulk discounts

Transmutation: the best superpower in the X-men universe

V

Valence: how much love an atom can spread around

Valence Electron: the swinger of the subatomic particles

Viscosity: think of comparing honey to water and how runny it is

W

Wave Function: what surfers try to figure out
Weak Acid: an acid that needs to go to the gym
Weak Base: a base that needs to go the gym

X

X-rays: something so safe that the people who work with it must wear lead aprons.

Y

Ytterbium: when someone asks if they bite someone

Niels Bohr Misquote: "Unconventional Publishing has a free adult joke for you to download on their website. Just head over to "www.unconventionalpublishing.com.au"

Arthors Note

Hi my friends, welcome to *The Fantastic Book of Chemistry Jokes, Authors Edition.* This book has been niggling at me for a long time, upon receiving some advice, I decided to initially write a smaller book. I felt I had more stories to tell, more jokes to give, hence the revised edition. I sincerely hope you had as many laughs as I did writing this book. If you did enjoyed reading this, please leave a review, it would be greatly appreciated, and would help me out immensely.

**Free Adult Jokes Book **

You can also find another free joke book on the company's website, one that is not for sale, one that is not for the faint heart, one that can be considered a bit risqué and politically incorrect. Consider this a gift for taking the time to purchase and read my book but be warned, only download it if you are not easily offended.

www.unconventionalpublishing.com.au

If you want to see any other professions being roasted or want a particular joke on a shirt, please visit our website and let us know.

Kind Regards

Shane Van